I0490328

Table of Contents

Cannabis Dosage Guide For

Pain,Anxiety And General Health

CANNABIS DOSAGE GUIDE

Dosage Guide On How To Use Cannabis For Chronic Pain,Anxiety,Epilepsy And General Health.

BRUTE WES

Copyright@2020 By Brute

Wes

All rights reserved.No part of this publication may be reproduced,distributed or transmitted in any form or by any means including photocopying,recording or other electronic or mechanical methods without the prior written permission of the publisher,except in the case of brief quotations embodied in critical reviews and certain other non commercial uses permitted by copyright law.

ISBN 978-1-67812-349-9

90000

9 781678 123499

INTRODUCTION

The greater part of U.S. states and the District of Columbia have sanctioned medical marijuana in some structures and more are thinking about bills to do likewise.

However,while numerous individuals are utilizing marijuana,the FDA has just endorsed it for treatment of two uncommon and serious types of epilepsy, Dravet disorder and Lennox-Gastaut disorder.Medical marijuana utilizes the marijuana plant or synthetic substances in it to treat sicknesses or conditions.It's fundamentally a similar item as recreational cannabis,however it's taken for medicinal purposes.

The marijuana plant contains in excess of 100 unique synthetic substances called cannabinoids.Every one differently affects the body. Delta-9-tetrahydrocannabinol (THC) and cannabidiol are the principle synthetics utilized in medication.THC additionally delivers the "high" individuals feel when they smoke weed or eat food containing it.

CHAPTER ONE

Cannabis And Marijuana Pharmacy

Today you will gain proficiency with the potential advantages of marijuana. The vast majority of the potential advantages of marijuana are synonymous with the medical advantages that the cannabinoids inside marijuana like THC and CBD give.

Be that as it may, weed authorization particularly contrasted with marijuana use has more potential advantages than well being related advantages, as you'll learn today. A discussion on the advantages of marijuana,in any case,wouldn't finished without featuring the potential unfavorable health impacts of cannabis too.

Despite the fact that marijuana has numerous potential advantages, you'll likewise discover that cannabis use has genuine potential unfavorable health impacts and you ought to be mindful so as not to mishandle it.

Here Are Some Of Therapeutic Benefit Of Marijuana

Cannabis animates craving and weight gain in AIDS,cancer and anorexia nervosa patients

Numerous examinations during the time have indicated cannabis use is straightforwardly connected with extended craving and weight. The accurate system is as yet vague, however we do know there's a high centralization of cannabinoid receptors in brain parts that are related with control of nourishment intake.

Marijuana diminishes spasticity related with various sclerosis (MS)

The impacts of weed on psycho motor brokenness and spasticity related with numerous sclerosis are very much recorded.

Numerous sclerosis patients accepting deficient alleviation from customary medications, in various examinations, have detailed an emotional reduction in spasticity when utilizing marijuana.

Particularly smoking weed has solid proof for diminishing spasticity and pain.

Marijuana could help with amyotrophic lateral sclerosis (ALS)

ALS is one of the 3 most savage neuro degenerative ailments known to man. And keeping in mind that there's a ton of research to be done, the underlying examinations done on animal show that cannabinoids have numerous neuro protective properties which can be useful in treating ALS. Weed reliefs all signs related with spinal damage in certain patients

Spinal damage can be a significant condition that meddles with your day by day life. A portion of the indications which especially can meddle with your general wellness are pain, muscle spasms,spasticity and troubles resting or sleeping.

Pot Or Marijuana Can Help With Epilepsy

An ever increasing number of studies recommend that epileptic movement is identified with changes in the levels and dispersion of cannabinoid receptors in the hippocampus.

So it's no sudden that for over decades, cannabinoids have been appeared to decrease seizures successfully, particularly in patients that have resistance from conventional medications.

Weed Could Reduce Constant Torment

Constant torment is an unpredictable condition which includes numerous elements and is a reason for decreased personal satisfaction.

Different clinical examinations have demonstrated that smoked or disintegrated cannabis can prompt a huge lessening in constant neuropathic torment related with different disorders and infections.

Pot also diminishes cerebral pains and headache assaults in certain patients.Ask any individual who is experiencing overwhelming headache and that individual will mention to you what a hellfire he/she experiences while a headache assault erupts.

12

I have 2 companions who experience the ill effects of headache every once in a while and pot causes them hugely in adapting to it. Cannabis could help with osteoarthritis (OA) by forestalling ligament breakdown

Pot can also help with all indications related with fibromyalgia

Fibromyalgia is an interminable torment condition which can seriously influence your day by day life through weariness, intellectual and enthusiastic aggravations.

Albeit nothing recommends that cannabis could fix fibromyalgia, a developing proof shows that marijuana could help with treating its signs and symptoms.

Weed can improve indications related with dystonia up to half

Dystonia is a neurological development issue which causes unusual developments and muscle contractions .

While there's still a great deal of research to be done, an examination done right in 1986 found that day by day admission of 100mg to 600mg of the cannabinoid CBD co administered with standard drug, improved dystonia up to half.

This is a significant and gigantic improvement, and as I would see it, warrants a more profound investigation of the potential advantages that cannabis has for dystonia patients.

Marijuana could debilitate the movement of Huntington's infection (HD)

Marijuana also assists with the development issue related with Parkinson's infection (PD) and development issue brought about by the utilization of its customary prescriptions

Pot assists with Tourette's disorder

Weed could help with glaucoma

Cannabis goes about as a bronchial dilator in asthma patients

Cannabis can bring down blood vessel circulatory strain in patients with hypertension

Marijuana can be a viable stimulant

Marijuana can help with rest issue by successfully improving rest quality

Marijuana can be utilized to control the maltreatment of progressively hurtful medications

CHAPTER TWO

Everything About Medical Marijuana

Therapeutic marijuana additionally called medicinal cannabis is a term for derivative of the Cannabis sativa plant that are utilized to calm severe and chronic symptoms.

Is Therapeutic Marijuana Lawful In The U.S.?

U.S. government law precludes the utilization of entire plant Cannabis sativa or its subsidiaries for any reason. CBD got from the hemp plant (< 0.3% THC) is legitimate under government law to devour.

Numerous states permit THC use for medicinal purposes.Government law regulating marijuana overrides state laws. Along these lines,individuals may even now be captured and accused of ownership in states where weed for therapeutic use is legitimate.

When Is Medicinal Marijuana Suitable?

Studies report that medicinal cannabis has conceivable advantage for a few conditions.State laws change in which conditions qualify individuals for treatment in regards to medicinal marijuana.In case you're thinking about marijuana for medicinal use check your state's guidelines and rules.

Contingent upon the state, you may fit the bill for treatment with medical cannabis on the off chance that you meet certain prerequisites and have a passing condition, for example;

Alzheimer's ailment

Amyotrophic Lateral sclerosis (ALS)

HIV/AIDS

Cancer

Crohn's ailment

Epilepsy and seizures

Glaucoma

Different sclerosis and muscle Spasms

Serious and constant torment

Serious Nausea

On the off chance that you are encountering awkward indications or negative reactions of therapeutic treatment, particularly agony and sickness converse with your doctor pretty much about the entirety of your alternatives before attempting marijuana.Specialists may consider medical cannabis as an alternative if different medicines haven't made a difference.

Is Therapeutic Cannabis Safe?

Further investigation is expected to respond to this inquiry,yet conceivable negative side reactions of medicinal cannabis may include:

Rapid Increase in heart rate

Tipsiness

Hindered focus and memory

More slow response times

Expanded danger of heart attack and stroke

Increased In Appetite

Potential for Addiction

Cyclic vomiting disorder

Mental illness

Signs Of Withdrawal

Is Therapeutic Marijuana Accessible As A Doctor Prescribed Medicine?

The U.S.A FDA has endorsed one cannabis-inferred and three cannabis-related medications namely dronabinol, nabilone and cannabidiol.

Dronabinol and nabilone can be recommended for the treatment of nausea and vomiting brought about by chemotherapy and for the treatment of anorexia related with weight reduction in individuals with AIDS.Cannabidiol can be endorsed for treatment of extreme types of childhood epilepsy.

What You Can Anticipate

Therapeutic marijuana may be found in different forms,including:

Oil for vaporizing

Pills

Topical applications

Oral mixture

Dried leaves and buds

How and where you buy these substances lawfully differs among the states that permit medicinal utilization of cannabis. When you have the item, you regulate it yourself. How regularly you use it relies upon its forms and your signs. Your signs may help and symptoms likewise will differ dependent on which type you are utilizing.

The snappiest impacts happen with inward breath of the vaporized form. The slowest begins with the one in form of pill. Some medicinal marijuana is produce to give symptoms alleviation without the inebriating, state of mind adjusting impacts related with recreational utilization of cannabis.

CHAPTER THREE

How Do You Get Medical Marijuana And Method Of Usage?

How Can Marijuana Help?

Cannabinoids the potent dynamic synthetic substances in therapeutic marijuana are similar to the chemicals the body makes that are engaged with craving food, memory,movements and pain.Research recommends cannabinoids may:

Diminish anxiety

Relieve Pain

Control nausea and vomiting brought about by chemotherapy

Destroy cancer cells and slow down tumor development

Loosen up tight muscles in individuals with MS

Invigorate appetite and improve weight gain in individuals wit AIDS

Will Medicinal Marijuana Help With Seizure Issue?

Medicinal marijuana got a great deal of consideration a couple of yeas back when guardians said that a unique type of the medication helped control seizures in their kids.

The FDA as of late affirmed Epidiolex,which is produced using CBD as a treatment for individuals with serious or difficulty in treating seizures.In regards to study few people had a sensational drop in seizures in the wake of taking this medication.

Which States Permit Medicinal Marijuana?

Medicinal marijuana is lawful in 33 states in United State and the District of Columbia:

The Frozen North

Arizona

Arkansas

California

Colorado

Connecticut

Delaware

Region of Columbia

Florida

Hawaii

Illinois

Louisiana

Maine

Maryland

Massachusetts

Michigan

Minnesota

Missouri

Montana

Nevada

New Hampshire

New Jersey

New Mexico

New York

North Dakota

Ohio

Oklahoma

Oregon

Pennsylvania

Rhode Island

Utah

Vermont

Washington

West Virginia

States that permit limited usage include: Alabama, Georgia, Iowa,KentuckyMississippi, Missouri, North Carolina,South Carolina, Virginia, Wisconsin and Wyoming.

How Would You Get Medicinal Marijuana?

To get medicinal marijuana,you need a well written proposal from an authorized specialist in states where it is legitimate.Few out of every doctor or specialist is eager to prescribe medical marijuana for their patients.

You should have a condition that fits the bill for therapeutic cannabis use.Each state has its very own rundown of qualifying conditions.

Your state may likewise expect you to get a medicinal marijuana ID card.When you have that card,you can purchase at a medicinal marijuana store known as dispensary.

How Would You Take Marijuana?

To take therapeutic cannabis,you can:

Smoke it

Breathe it through a gadget considered a vaporizer that transforms it into a mist

Eat it for instance, in a brownie or candy

Apply it to your skin in a lotion,oil or cream

Put a couple of drops of a fluid under your tongue

CHAPTER FOUR

Cannabis Dosage Guide For Pain,Anxiety And General Health

There are no official dosage guide identified with effective CBD use. Everything we can do is take a look at CBD portions utilized in examines, investigate recounted reports at our own encounters with CBD oil.

It's likewise essential to take note of that in light of the fact that various strategies for CBD oil utilization accompany various degrees of bioavailability, a successful measurements is additionally profoundly subject to the technique for use.

Add to that the way there are huge contrasts between people in the use of CBD and other physical attributes like weight and body mass, and you'll rapidly comprehend why it's hard to give a general measurement rule for powerful CBD oil usage.

The greatest portion of CBD ever taken in an examination, with no genuine reactions is 1500mg of CBD. Taking 1500mg of CBD implies freeing a bottle of CBD oil in a solitary use.

CBD oil isn't intended to be taken in that manner, for the most part, you drop just 5-10 drops of CBD oil under your tongue, never will anybody prescribe completing a major bottle of CBD oil in a solitary swallow.

CBD Dose For Epilepsy

One research that took a look at the impacts of CBD on patients with epilepsy found that 200-300mg of CBD use every day had a positive results on the event of convulsive emergencies in 7 of 8 patients.

The normal day by day portion detailed by the patients for refined CBD was 25.3 mg/kg/day, while the normal day by day portion of CBD proportional announced for CBD-rich Cannabis extricate was just 6.0 mg/kg/day.

Obviously, it's still very costly to take CBD at an every day portion of 6.0 mg/kg/day as a CBD-rich Cannabis extract, yet it's significantly less expensive than taking CBD at a day to day portion of 25.3 mg/kg/day as a cleansed CBD item

35

like CBD Isolate. Further investigations should show whether you can accomplish the equivalent gainful outcomes for different conditions like anxiety and pain with 4 times less CBD when taken as CBD-rich Cannabis extricate contrasted with when taken as a filtered or purified CBD item.

CBD Dose For Psychosis

There haven't been any huge scale research that took a gander at the impacts of CBD on psychosis in people. Also, the couple of studies that took a gander at the impacts of CBD on psychosis discovered constrained proof for its utilization.

For instance, in a situation study, a 19-year elderly person with schizophrenia got a fruitful treatment with a portion of 1200 mg/day of purified CBD. In another investigation that looked at the impacts of CBD versus a customary antipsychotic named amisulpride,scientists found that a day by day portion of 200mg CBD, expanded stepwise by 200mg every day to a day by day portion of 200 mg multiple times daily suming it up to 800 mg every day each within the first week, was as successful as the conventional anti psychotic prescription amisulpride, while CBD had an unrivaled symptom profile that is less negative side reactions.

Once more, these dosages of CBD are high and except if your insurance agency covers your CBD buys, it may be monetarily difficult to supply yourself with these amount of CBD.

CBD Oil Dose For Anxiety

A recent report that inspected the consequences of 25+ human trial, clinical, epidemiological and constant examinations found that the present proof firmly underpins the potential for CBD as a treatment for anxiety issue at oral dosages going from 300mg to 600mg. There's particularly solid proof for CBD decreasing tentatively induced anxiety, similar to open speaking tests, and for lessening social anxiety issue.

Essential to note here is that the respondents utilized a purified type of CBD and not a full-plant extract.This implies,shockingly,no portion recommendation can be given concerning taking CBD as a full-plant extract for tension and anxiety.

Yet, in the event that the examination that thought about the adequacy of full-plant extricate CBD items versus purified CBD items in epilepsy patients is in any capacity characteristic for anxiety too (a full-plant remove CBD item requiring 4-times less of CBD contrasted with a cleansed CBD item for similar impacts), the prescribed portion of full-plant separate CBD oil to treat anxiety would be identical to 75mg to 150mg of CBD.

CBD Dose For Pain

Most examinations that discovered painkilling impacts with cannabis-based medications utilized a mix of both THC and CBD.

Lamentably,the dosages utilized in these examinations can accordingly not be translated for a CBD dosage.

There is also an ongoing examination that discovered painkilling impacts of CBD in kidney transplant patients, with an underlying portion of 100 mg/day and a dynamic increment up to 300 mg/day spread out throughout the day in little dosage.

A fascinating note here is that this equivalent examination found that more CBD isn't in every case better. One of the patients experienced more grounded painkilling impacts with lower dosages of CBD it's uncertain how much lower than 300 mg/d.

CBD Dose For General Health

Research that take a gander at the impacts of CBD consistently take a gander at the impacts of CBD on a particular condition or the impacts of CBD on the symptoms of a particular health condition.

So,CBD has solid calming oxidant attributes. You needn't bother with a particular condition to receive the rewards of

CBD oil. Anybody can profit from food with solid calming

and cancer prevention properties.

Our informal suggestion for taking CBD oil for general

wellness is 10-20mg of CBD every day, as a full-plant

extricate/full-range CBD oil.

www.ingramcontent.com/pod-product-compliance
Lightning Source LLC
Chambersburg PA
CBHW031504210526
45463CB00003B/1078

* 9 7 8 1 6 7 8 1 2 3 4 9 9 *